John W. Thrailkill

An Essay on the Causes of Infant Mortality

being a brief account of the origin of the feebleness and diseases which

afflict and destroy so many children under five years of age

John W. Thrailkill

An Essay on the Causes of Infant Mortality
*being a brief account of the origin of the feebleness and diseases which afflict and
destroy so many children under five years of age*

ISBN/EAN: 9783337853075

Printed in Europe, USA, Canada, Australia, Japan

Cover: Foto ©berggeist007 / pixelio.de

More available books at **www.hansebooks.com**

ON THE

Causes of Infant Mortality;

Being a brief account of the origin of the feebleness
and diseases which afflict and destroy so many
children under five years of age.

— BY —

JOHN W. THRAILKILL, M. D.,

St. Louis, Mo.

———>•◆•———

ST. LOUIS:
S. W. BOOK AND PUBLISHING CO., 510 WASHINGTON AVENUE.
1869.

PREFACE.

The articles of which this little work is composed originally appeared in the Sunday edition of the *Missouri Republican*. The author had no notion at the time they were written of ever putting them into the form of a book, but having been earnestly solicited to do so by numerous parties who read them as they first came out, at length consented, and now begs leave to present them to the public in this form. Some additions and corrections have been made to the articles, and they are submitted without an apology for the numeros errors and imperfections with which they still abound.

The articles are so brief as to be little more than an index to the great subject upon which they treat. But it is hoped that they may serve to stimulate thought and investigation in a quarter where they are most needed, and will do most good, namely—among parents. There are many parents who would willingly introduce reforms into the prevailing modes of managing their children if they could be convinced that there are better and more successful modes of treatment than have been handed down to them from their parents. But the power of education is so great in its influence over the conduct of individuals and nations that it is next to an impossibility to convince people that any other way is better than the one they have been taught. The only way to introduce any given reform, in opposition to this prejudice of education, is to continually " agitate the question "—" keep it before the people." All reforms have been introduced in this way, and printers' ink is by all means the most successful mode of accomplishing this end. It is by the power of the PRINTING PRESS that the thoughts of the best and wisest of mankind are rapidly becoming the common property of all.

THE AUTHOR.

No. 312 N. Sixth Street, St. Louis, Nov., 1869.

CONTENTS.

Preface..Page 4

ARTICLE I.

Hereditary Causes of Infant Mortality................   5

ARTICLE II.

Causes of Infant Mortality arising with the mother during Gestation.....................................   13

ARTICLE III.

The Extrinsic or Exciting Causes of Infant Mortality....   21

ARTICLE IV.

Causes of Infant Mortality arising from the Mismanagement of Parents and Nurses.........................   29

ARTICLE V.

Infant Mortality arising from such sources as should
come within the Province of the Public Authorities..   40

ARTICLE VI.

Causes of Infant Mortality arising from the Mismanagement of Physicians................................   49

A Treatise on the Physiological Management of Infancy
and Childhood.....................................   61

AN ESSAY

ON THE

CAUSES OF INFANT MORTALITY.

ARTICLE I.

I noticed in the *Republican* of the 8th inst. "A Challenge to the Medical Faculty," by Chas. Miller, F. H. Gerhold and others. The principal proposition contained in this "challenge" is a demand upon the medical faculty to explain "the wherefore of such an alarming death-rate among children, and why the boasted science of the medical profession does or can do nothing." Such a demand on the part of the people is perfectly rational and legitimate, and should be responded to by the profession with that alacrity which has ever distinguished them as the philanthropic conservators of the public weal. The importance of the subject can not be over-estimated. It involves the dearest interests of the race, and demands the attentive and earnest consideration of all. The limits of a newspaper article will permit nothing but the briefest possible consideration of the subject; hence I shall enter at once upon the main topics without further preliminaries.

The statement is made in the above mentioned challenge "that about forty per cent. of all the deaths are children under five years." It would be foreign to my purpose either to substantiate or call in question the truth of this statement. Besides, such an inquiry would be barren of practical results, because its approximation to the truth will be admitted by all intelligent observers.

It can not be possible that so large an infant mortality is the order of Providence, and that its causes are beyond the control of society. Such a view would necessarily lead to fatalism—a doctrine refuted by the every-day experience of all. If this be a scheme of Providence, why trouble ourselves about His affairs? Why believe in the utility of sanitary arrangements or the efficacy of remedial measures? If the Divine Being has the matter under his immediate supervision, and it is his will that so large a portion of the race shall perish in infancy, man is thereby relieved of all responsibility in the matter. Not only this, but he is deprived of all power to modify the result, either for the better or worse, and his attempt to do so is an impious effort to act in defiance of the will of the Creator. If this be the order of Providence it is obviously our duty to piously close our eyes, fold our arms, and be resigned to His will. But if, on the other hand, man is a free agent and an accountable being, responsible for the manner in which he uses the talents intrusted to his care, and capable of modifying the conditions of nature with which he is surrounded for the promotion of his own happiness and that of his fellow-creatures, then it must be evident to all that the well-being of infancy depends, in a great measure, upon the treatment it receives at his hands, and that he becomes accountable for all the suffering arising from his ignorance and mismanagement. Whether man is *morally* culpable for pursuing a wrong mode of treatment because he is ignorant that it is such, is a question I shall leave for philosophers to decide. But the physical suffering of the young being arising from such mismanagement is precisely the same, whether it arise from ignorance or a willful infliction of those having charge of it. For example: If a parent administer to her child any given drug which she believes to be innocent, but which proves to be poisonous to it, its sufferings are none the less on account of her ignorance, but precisely the same as if she had given it with intent to destroy its life.

No intelligent being will for a moment contend that sickness and death can be wholly banished from infancy by any system of management which man, in his present state of enlightenment, can adopt; for the seeds of early decay are often sown by parents even of the third and fourth generations, and they will necessarily, in some instances, ripen into a harvest of disease and death, despite the most favorable influences and the best treatment. Besides, accidental infringements of the laws of health will necessarily occur sometimes to the most enlightened and experienced individuals.

"The successful rearing of every living being," says Dr. Combe, "depends chiefly on the proper adaptation of its treatment to the laws of its constitution. Where these are in harmony, the failures will be few and unimportant, and arise chiefly from those unavoidable accidents and exposures to which all created beings are, and will continue to be, more or less subjected. But where the treatment and laws are not in harmony, failure, disease and ultimate death may be expected as the most frequent and certain results."

The causes which affect the health and life of infancy will be considered under three distinct heads, viz.:

First, HEREDITARY, or those constitutional peculiarities derived from both parents, and supposed to have their origin at the time of conception.

Secondly, MATERNAL, or those derived from the mother during gestation.

Thirdly, EXTRINSIC, or those acting on the child from without, after its birth.

HEREDITARY CAUSES.

Hereditary causes play a very important part in the production of infant mortality. Puny parents, or those weak in vitality, can not beget vigorous offspring. "Like begets like." No fact is more patent than that many children are born so feeble in constitution as to be swept out of existence by the slightest causes. Every physician of experience must have noticed how difficult it is to restore the sick children of some parents, compared with others, and how they often perish from causes so slight as to make no impression on others more favored in hereditary constitution. Indeed, every judicious physician forms his prognosis of a sick child, in a great measure, from the temperaments or physical aspect of the parents.

The various conditions of parents affecting the health and vigor of offspring are:

" *First.* Natural infirmities of constitution derived from their own parents.

" *Secondly.* Premature marriages, especially of delicate females, and persons strongly predisposed to hereditary disease.

" *Thirdly.* Marriages between parties too nearly allied in blood, particularly where either of them is descended from an unhealthy race.

" *Fourthly.* Great disproportion in age between the parents.

" *Fifthly.* The state of the parents at the time of conception."*

Sixthly. Incompatibility of temperaments between the parents.

These may seem a numerous and formidable array of disqualifications in parents to beget healthy offspring; nevertheless, they each contribute in a very important degree to the production of that delicacy and feebleness of constitution so extensively prevalent everywhere amongst children, which render them so susceptible of every extrinsic source of disease, and so unable to bear up under its ravages. The transmission

---

*Combe on " Infancy," p 55, *et seq.*

from parent to child of the tendency to the more obvious forms of constitutional disease, such as consumption, scrofula, insanity, &c., is universally recognized. Indeed, the fact is so patent, and the knowledge of it so universal, that all mankind must have known it without the aid of revelation. No one afflicted with any one of these diseases should ever become a parent; and those in whom such constitutional *tendency* exists should select healthy partners of a different temperament.

If the tendency to any given disease exists in *both* parents, it will be doubled in the offspring, and they will be likely to fall an early prey to it. It is in this way that consumption has become so prevalent and fatal amongst us. The seeds of death are thus continually sown by parents, as tares among wheat, and they come forth and yield an abundant harvest. The spare, thin-breasted youth falls in love with the hectic flush on the delicate maiden's cheek, and they become united into one being, and their children, one after another, descend into the grave before they are able to call their parents' name; or if they live to grow up and propagate their kind, it is but to sow the seeds of disease and death and people the graveyard with little graves.

Early marriages, before the parties have attained complete growth, is a frequent source of delicacy and disease in infancy. It is often the case that the first children of such parents are feeble, decay and die, while others, begotten after maturity is attained. are healthy. Besides, early marriages are one of the most prolific sources of early decay and premature old age in our American women. A girl married at fifteen will be older in physical decay. health and beauty at thirty than one married at twenty will be at forty-five, *cæteris paribus.*

Great disparity of years in the parents is often a source of imperfection in offspring. The offspring of old men and young wives rarely attain great age. and are frequently subject to lingering infirmities.

The state of parental health at the time of conception has a greater influence on the future offspring than is generally imagined. It is doubtless the case that many children owe their constitutional infirmities to some acquired or temporary infirmity of one or both parents existing at the time of conception. Quite a number of well authenticated cases of idiocy and insanity have been reported as having occurred in children as the result of a drunken debauch on the part of generally temperate fathers; and if the unwritten records of sin could be traced they would doubtless exhibit many cases of the kind.

"A stronger motive to regularity of living, and moderation in passion," says Dr. Combe, "can scarcely be presented to a right-minded parent than the simple statement of their permanent influence on his future offspring. Many a father has grieved over, and perhaps resented, the distressing and irreclaimable follies of a wayward

son without suspecting that they actually derived
their origin from some forgotten irregularity of
his own." Children begotten by parents while in
a state of debility from great mental anxiety,
from cares and troubles, or some harassing bodily
ailment, are generally peevish, irritable, ill-na-
tured, and very susceptible of every extrinsic
source of disease. Such children are very liable
to convulsions—one of the most frequent and fearful sources of infant mortality. But this is only
one of a legion of ills arising from the same
source.

The source of much infant mortality has its origin in *incompatibility of temperament in the parents.* Prof. W. Byrd Powell deserves the lasting
gratitude of mankind for discovering and eluci-
dating this fact. I know of many instances of
man and wife so incompatible as to be either
wholly unable to reproduce living children, or if
they are viable at all, they live to an inheritance
of disease and suffering. while the parents themselves are sound in constitution and enjoy good
health. Two persons, both of whom have a *lymphatic* temperament. should never become man
and wife if they desire healthy offspring. If the
temperament be strongly marked the chances are
that they will have no living children; but if it
be partial they may live for a time. The same
injunction is applicable to those having a cephalic
or nervous temperament. The character of the
diseases, however, to which the offspring will be
subject, will vary somewhat as does the tempera-

ment of the parents. Lymphatic parents generally produce scrofulous children, and those in whom the nervous temperament predominates will produce those more subject to disease of the brain and nervous system. Persons possessed of either of these temperaments in a marked degree should select partners of a different one if they desire viable and healthy offspring.

My next paper will be on the causes of infant mortality originating with the mother during gestation.

ARTICLE II.

MATERNAL CAUSES, OR THOSE ARISING WITH THE MOTHER DURING GESTATION.

Having briefly considered in my last paper the HEREDITARY causes of infant mortality, the second division of the subject—the MATERNAL, or those brought to bear on the child through the maternal system during gestation—demand consideration, as scarcely less important. Upon the mother, much more than the father, depend the health and vigor of offspring. The ancient Spartans recognized the importance of this fact, and bestowed especial pains on the education of their daughters. When a foreign woman said to Gorgo, the wife of Leonidas, " You Spartan women are the only women in the world who rule the men," Gorgo replied, " *We are the only women that bring forth men.*" Lycurgus, their great

law-giver, believed the education of the youth to
be the most glorious work of the rulers of the
State. He began, accordingly, by regulating marriages with reference to physical adaptability, in
order to insure to the coming generation a healthy
physical organization as a solid foundation upon
which to build the prosperity of the State. "He
ordered the virgins to exercise themselves in running, wrestling and throwing quoits and darts ;
that their bodies being strong and vigorous, the
children afterward produced from them might be
the same, and that, thus fortified by exercise, they
might the better support the pangs of child-birth
and be delivered with safety."

The health and character of the future offspring
are greatly influenced by the conduct of the
mother during carriage. Great numbers of children die from causes originating during this time.
One of the most striking instances on record illus-
trating the truth of this fact is recorded in the
Dictionnaire des Sciences Medicales, on the au-
thority of Baron Percy, an eminent French mili-
tary surgeon and professor, "as having occurred
after the siege of Landau, in 1793. In addition
to a violent cannonading, which kept the women
for some time in a constant state of alarm, the ar-
senal blew up with a terrific explosion, which few
could listen to with unshaken nerves. Out of
ninety-two children born in the district within a
few months afterward, Baron Percy states that
SIXTEEN died at the instant of birth ; THIRTY-
THREE languished for from eight to ten months,

and then died; EIGHT became *idiotic*, and died before the age of five years; and TWO came into the world with numerous fractures of the bones of the limbs, caused by the convulsive starts in the mother excited by the cannonading and explosion! Here, then, is a total of fifty-nine children out of ninety-two, or within a trifle of TWO out of every THREE, actually killed through the medium of the mother's alarm and its natural consequences upon her own organization." This is, indeed, a striking example, and perhaps but few such occur; but it forcibly illustrates the influence of the mental emotions of the mother upon the unborn child, and suggests the necessity of avoiding every cause calculated to disturb the equilibrium of her mind. Doubtless many children perish, or are rendered invalids for life, by domestic bombardments, cannonadings and explosions, confined within the four walls of home and unknown to the outside world. Connubial infelicity, social discord, misfortunes and the various accidents to life and limb to which all are more or less liable, frequently subject the prospective mother to bitter griefs, heart-rending anxieties, terrible frights and wild alarms, all of which have a deleterious influence on the developing infant. It is highly probable that all violations of the laws of health affecting the mother during this time produce similar effects on the young being she carries; disturbing its healthy development or entailing on it lasting infirmities, according to the magnitude or persistence of the parental offense. Although the evil effects of such parental infringements may not be obviously manifested in the infant after its birth, we should not, on that account, conclude that they have had no deleterious effect; for the constitutional vigor of the infant may be impaired, and doubtless is, in many cases, in an imperceptible degree, yet sufficient to render it less capable of resisting the extrinsic sources of disease, and less able to withstand its attacks.

There are many women, alas! *very* many, who live in habitual disobedience to the laws of health. A vast majority of them do so from sheer ignorance, never having been educated in that most important of all sciences— physiology. On account of this general ignorance, the habits and customs of society are not formed and sustained from any just conception of their appropriateness to the preservation of life. health and longevity; for these things are rarely thought of until they become imperiled. Many of the most detrimental habits of society are conceived in ignorance, brought forth in fashion, and nurtured in indolence and dissipation. Women, in the condition of which I am treating, are frequent sufferers from these *unphysiological* habits, some of them being peculiar to the sex. While in this condition she is more susceptible of the evil consequences of bad habits; and she becomes sickly from causes, the influence of which she is able at other times to resist without any marked deterioration of her health. This is due to the fact that all the energies of her being are required at this time in the development of her offspring, and none are left with which to combat the evil effects of bad habits.

I will now enumerate and briefly discuss a few of the most common infringements of the laws of health—those from which women, during gestation, most frequently suffer, and bring consequent disease upon their offspring, namely :

Irregularity in habits of eating and sleeping. This is more frequently exemplified among the rich and fashionable, the opera-goers, and those inordinate lovers of admiration who make and visit parties and fashionable gatherings in order to feast their vanity upon the flattery of fools. See the numbers who nightly visit places of amusement, witness exciting performances, return home at midnight, eat heavy, indigestible lunches, and perhaps drink wine, and then retire to bed to a restless, dreamy, feverish slumber, continued well into the next day

Prospective mothers often partake of too great a variety of highly seasoned and stimulating food. Many keep up a morbid excitement of the imagination by reading brain-sick novels. Want of a sufficient amount of invigorating exercise in the open air is a prolific source of ill health in this class of women, especially in cities. Living in tight, ill-ventilated and often over-heated rooms. Inattention to cleanliness and the bath. Among the poor, over-work, want of sufficient amount of nourishing food, and the cares, troubles and anxieties of life. And last, but not least, that great

lever of the physical degeneracy of the race—
tight lacing. Philanthropists, physicians and
physiologists have been preaching, lecturing,
writing and praying for the last fifty years to set
forth the enormity of this suicidal, matricidal, in-
fant-destroying sin ; but their warnings fall upon
the ears of society as sounding brass and a tink-
ling cymbal, and the practice still continues una-
bated. Girls, before they have entered their teens,
must have their waists screwed up in an immov-
able vise of steel, whalebone and lacings, to sat-
isfy the imperious demands of a blighting fashion,
created and sustained by depraved and distorted
views of the divine beauty of the female form.
Thus educated from childhood, women grow up
and depart not from the teaching, but instruct
their children to do likewise ; and thus the evil
has been propagated from generation to generation, each having to suffer for the sins of previous
ones as well as its own, until the cumulus upon
the present one has become as a great, huge mill-
stone, bearing down and crushing the vitals out
of the race. No one but the experienced physician knows what a vast per cent. of the present
generation of women is afflicted with some disease or weakness originating from this cause,
which incapacitates them for the healthy perform-
ance of the maternal function. It is a habit which
thousands practice without a suspicion of the evils
it engenders. Very few, indeed, can be induced
to believe that the maladies they suffer can have
their origin in a habit apparently so indispensable

to their comfort. It is a habit, like most other
bad ones, which, when once adopted, can not be
discontinued without the utmost effort on the part
of its votaries. It is frequently begun in girlhood
or early womanhood; the compression used being
but slight at first, a feeling of comfort and support is experienced, and the carriage and personal
bearing are improved. Feeling pleased with the
experiment, it is continued. As time passes the
bones and muscles of the parts involved become
enfeebled instead of strengthened by the artificial
support, and the compression requires to be gradually increased in order to produce the same feeling of comfort and support it did at first. In this
way a degree of compression is finally attained
which at first the most heroic fortitude could not
have withstood. Yet it is continued, and deemed
an indispensable to comfort, while it is squeezing
the very life out of its victim. This, it will be
seen, is but a parallel of all other bad habits.
Like the fabled syrens, it allures to destruction,
and even the lute of Orpheus has never been able
to break the spell. The functions of the lungs,
of the heart, of the stomach, the liver, the bow-
els, the womb and the mammary glands are all
encroached upon and their healthy action im-
peded. Hence lung diseases, heart diseases, dys-
pepsia, liver complaints, bowel derangements and
womb complaints are all engendered by this un-
natural fashion. Under its blighting influence,
beauty, health, grace, cheerfulness and good tem-
per often become materially impaired before that

age is attained which fits woman for becoming a wife and a mother. Even the lilies fade where the roses should bloom, and the painter's art is called in to inspire dead beauty with life.

If the practice of lacing and wearing corsets is so perilous to female health at all times, it is but reasonable to suppose that it is much more so during gestation, and that the unborn infant must share the penalties of the sin. Doubtless every experienced physician has observed the great number of miscarriages, abortions, still-born and puny children among mothers who suffer from this practice. "The Romans were so well aware of the mischief caused by compression of the waist during gestation that they enacted a positive law against it. Lycurgus, with the same view, is said to have ordained a law compelling pregnant women to wear very wide and loose clothing." It is a well known fact that many fashionable ladies continue to wear corsets and lacings in order to preserve their wonted personal appearance in society until gestation is well advanced, loosing them from time to time barely sufficient to give breathing room. Such a practice imperils the life and future health of the infant, to say nothing of the consequences to the mother.

It may be deemed indelicate and ungallant to refer to these matters in the manner I have done, but in summing up the causes of infant mortality such a course became unavoidable, as they deserve a prominent consideration, and therefore could not be passed over in silence. Those who desire enlightenment as to the causes of the evils befalling the infant population must not shrink from the investigation through false modesty.

My next paper will be on the extrinsic causes of infant mortality, or those operating on the child after birth.

ARTICLE III.

THE EXTRINSIC CAUSES OF DISEASE.

Having in my previous article treated of the causes of sickness and death among infants arising from parentage previous to birth, the little stranger, as an independent being, will now make his *debut* and demand our consideration. It will be seen, by recalling the facts presented in my former papers, that his existence previous to birth has been frequently sorely jeopardized by the constitutional infirmities, accidental misfortunes, or cruel wickedness of his parents. Is it surprising, considering the many constitutional infirmities and wicked violations of the laws of health under which the present generation of parents suffer, that so many children are born to an inheritance of physical and moral depravity, disease and death? Many come into the world as frail as a gossamer and are swept into the grave by the gentlest breath of heaven. Hence it is that between one-third and one half of all children born in civ-

2

ilized society die in five years.   Such has been the
case for centuries, or from as remote a date as
statistical records can be had.   It has even been
much greater in old countries in past times.   The
present infant mortality in our city, of forty
per cent. under five years, is, therefore, not extra-
ordinary, as all the cities of the civilized world
present but a trifling variation from these figures.

But all the sources of weakness and disease
which beset the young previous to birth, numerous as we have in our former papers seen them to
be, are not sufficient to account for so great a
death-rate amongst them.   We must, therefore,
seek a solution of the problem in the various external conditions with which infancy is surrounded
subsequent to birth.   These constitute the EX-
TRINSIC or EXCITING causes of disease.

The causes of disease which originate with the
parents and operate on the infant previous to
birth serve only to diminish its constitutional
vigor, its vital force, its powers of resistance,
rather than to implant actual disease in its or-
gans.   Very few children, indeed, are born with
actual disease existing in them.   It may be justly
said that, with very few exceptions, all children
are born in a state of health.   But the powers
possessed by different individuals of resisting the
extrinsic sources of disease and retaining that
state are just as various as the parental influences
from which they have derived their origin.   Hence,
if we can preserve even the feeblest infant from
the action of those causes from which disease

arises, and by judicious management preserve it
in that state of health in which it is born, the
chances are that we will succeed in rearing it to
maturity. A puny stalk of corn sprung from a
defective seed, if taken as soon as it comes forth,
may, by judicious culture, care and preservation,
be reared to maturity and bring forth lusty ears.
But without this judicious culture it will perish.
So it is with the puny child. If the treatment it
receives be in harmony with the laws of its constitution, the chances are that it will live and
thrive, and *vice versa*.

A vast majority of the diseases which afflict
and destroy infancy arise from *mismanagement*.
This arises chiefly from the *ignorance* of the man-
agers, be they parents, nurses, physicians, legislators or what not; all are responsible, to a greater
or less extent, as we shall hereafter see. It is
granted that a certain per cent. of the mortality
arises from causes over which man has little or no
control. These causes I will enumerate, viz. :

First.—HEREDITARY PREDISPOSITION. This man
has no power to control after the child is born.

Secondly.— UNAVOIDABLE ACCIDENTS. These
will occur under the most enlightened system of
management which it is possible to conceive, but
will be diminished in proportion to the advance
of knowledge.

Thirdly.—THOSE CONDITIONS OF NATURE BY
WHICH WE ARE SURROUNDED OVER WHICH MAN
HAS NO CONTROL. Such are the changes of the
seasons—rain, sunshine, storm, &c.

From these three sources arises all that per cent. of infant mortality over which man has no control. Now, let us examine each briefly in detail, and see if we can come to any conclusions as to the amount of death arising from each of them, and what per cent. of the whole infant mortality arises from the aggregate of them. We must not hope for mathematical accuracy, but it is believed that a fair approximation to the truth can be arrived at.

First.—I believe that the experience of physicians and the statistical records of mortality will justify the conclusion that the number of infant deaths arising strictly from *hereditary* causes, without the action of those *secondary* causes over which man holds a controlling influence, is very small. It is a fact well known to all that many puny children who have inherited a defective constitution are, by the adoption of a judicious physiological system of management, reared to maturity and enjoy reasonable health to a good, ripe age. Now, is it not reasonable to conclude that if the same judicious management were applied to *all* cases of the kind, that the death-rate arising strictly from hereditary causes would be reduced to a very small amount? I believe the premises will warrant the conclusion that it would not exceed *ten per cent.* of the entire infant mortality under five years.

Secondly.—The per cent. of infant mortality arising from unavoidable accidents is small. I believe the statistical records of mortality in this and other countries will justify me in placing it at not above *five per cent.*

Thirdly.—What per cent. of infant mortality arises from the conditions of nature with which we are surrounded over which man has no control? This is a question not so easily disposed of. It involves the whole philosophy of man—the nature of the human constitution and its relations to the external world.

The most rational system of the philosophy of man that has ever been promulgated to the world (see *Combe on the Constitution of Man*) goes far to establish the fact that the constitution of man is in perfect harmony with the conditions of nature with which we are surrounded; that disease and premature death arise from violation of that harmonious relationship; that if man would accommodate his conduct to these conditions of nature—in other words, would live in habitual obedience to the fixed and immutable laws of the Creator, he would live in the uninterrupted enjoyment of health from birth to old age (provided his hereditary predisposition should be favorable), and finally die with scarcely a pang or a regret. Who will say that the constitution of man is, by nature, imperfect? The same will accuse its Author of being a bad workman. Who will say that the system of external nature around us, so beautiful, so varied, so harmonious, so pleasing to the senses, so inspiring to the mind, is not in harmony with the nature of man, who obviously stands pre-eminent among created things in this sublunary world ! The same will impeach the Creator as an imperfect God. No! the existence of such an appalling amount of disease and premature death is an evidence that we violate by our own conduct the conditions of nature established by the Creator, upon which life, health and happiness depend. Every law of nature and every object with which we are surrounded were doubtless designed by the Creator to contribute to the well-being and happiness of man, and yet what law of nature is not violated and made the source of disease and death? It is by the abuse of those things designed by the Creator for our greatest happiness that we make ourselves most miserable.

Man can not change the order of nature. "He can not arrest the sun in its course so as to avert the wintry storms and cause perpetual spring to bloom around him; but by the proper exercise of his intelligence and corporeal energies he is able to foresee the approach of bleak skies and rude winds, and to rear, to build, to fabricate, and to store up provisions; and by availing himself of these resources and accommodating his conduct to the course of nature's laws, he is able to smile in safety beside the cheerful hearth when the elements maintain their fiercest wars abroad." Even if man had the power to change the seasons and create perpetual spring, who will say that such change would be desirable? It is this harmonious evolution of the seasons which furnishes man with some of his greatest delights, and prepares him for the highest order of his enjoyments. Nature, ever changing, like a grand, moving panorama before him, never ceases to engage his fancy and delight his eye. Variety is beauty, is life; sameness is dullness, insipidity, death. It is in those regions of the earth where the greatest variety of climate abounds that man attains his grandest proportions in intelligence and the arts of civilization. The center of civilization is confined to a narrow zone around the globe, embracing in the different seasons of the year almost all the climates of the world; while the inhabitants of tropical regions, where perpetual spring abounds, are dull, stupid creatures, mere intellectual automata. It can not be that the Creator, after having scattered such a profusion of blessings on man in the temperate zone, has constituted the changes of the seasons a necessary cause of disease and death among infants. That the mortality of infancy is greater during the heats of summer and the bleak, cold winds of winter is of itself no evidence that hot weather and cold are necessarily causes productive of this result; but should rather, in my opinion, be deemed an evidence that we have failed to discover the true nature of the relationship existing between the infant constitution and these seasons, and to regulate our management in harmony with it.

The subject of atmospheric vicissitudes and malarial influences as causes of infant mortality may be deemed appropriate for consideration here. But, as many of the observations I have made in reference to the changes of the seasons will likewise apply to that of atmospheric vicissitudes, I shall, therefore, dismiss it, lest I become tedious. As for malarial influences, they generally arise from local causes over which man holds an entire controlling power; the subject may, therefore, be more appropriately discussed elsewhere.

The various diseases of children arising from *contagion*, namely: measles, small pox, whooping cough, scarlet fever, &c., are supposed to depend on influences over which man has but a partial control. However, vaccination being almost a specific for small-pox. very few children should perish from it if due diligence be practiced.

Conceding that the changes of the elements and other natural influences over which man has no control will necessarily produce a share of the infant mortality, what per cent. of the death-rate shall be accounted to these causes? After a due consideration of the facts and inferences, based on statistics (which I have no room here to introduce), I have deemed *twenty per cent.* a fair and liberal estimate.

It will now be seen that I have accounted for but *thirty-five per cent.* of the present mortality of children under five years. Whence originates the remaining *sixty-five per cent.?* It is my honest conviction, formed from a long and careful study of the subject, that *sixty-five* out of every hundred, or nearly *two-thirds* of all children that die under five years, perish as a consequence of *mismanagement*, arising chiefly from ignorance of the known laws of physiology and hygiene. The question at once arises, who are so ignorant, and from whose mismanagement arises so appalling a result?

My next article will be devoted to answering this question, when I shall proceed with the investigation "without fear or favor or hope of reward," save the pleasure it gives me in being able to contribute to the dissemination of what I believe to be useful knowledge.

P. S. I have heard of some criticisms on my previous articles from members of the profession, to the effect that, the facts and conclusions are very true, but that the public should not be enlightened on these subjects; that such knowledge should be kept among the profession. I believe all good people will join me in the conviction that any physician who entertains such a sentiment as this is either so ignorant as to deserve pity or a dangerous enemy to the best interests of society. But I can not believe that many of them entertain so degrading a sentiment.

ARTICLE IV.

CAUSES OF INFANT MORTALITY ARISING FROM THE MISMANAGEMENT OF PARENTS AND NURSES.

The conclusion was formed in my last article that but *thirty-five* children out of every hundred that die under five years perish from natural causes over which man has little or no control, and that the remaining *sixty-five* die from *mismanagement, arising chiefly from ignorance of the known laws of physiology and hygiene.*

The responsibility for this mismanagement rests on society in general, but, for the sake of convenience in discussing the subject, the responsible parties may be divided into three classes, viz. :

First.—Parents and nurses.

Secondly.—Legislators, or those whose duty it is to make and execute sanitary laws.

Thirdly.—Physicians.

When I shall have discussed each of these classes my essay will end.

First.—The chief sources of infant mortality arise from the mismanagement of parents, and others under their control, viz.: nurses.

It must be obvious to every thinking mind that our chief hope of greatly reducing the present infant mortality lies in *preventing* the diseases instead of *curing* them; because, considering the many infirmities of constitution which children of the present generation derive from their parents, many of the diseases which afflict them will necessarily prove fatal, despite all the remedial influences of medicine, potent as they are. And it is also obvious that the prevention of sickness amongst children must necessarily be the work of parents, as such, more than all other parties put together. Unless the management of the parents be judicious many diseases will arise and often prove fatal, even with the best sanitary regulations on the part of public authorities and the most skillful treatment by physicians.

It has ever been the characteristic of ignorance and its twin sister, superstition, to attribute every phenomenon in nature, the cause of which can not be obviously seen, to the dispensations of Providence. This view has had its influence on society to such an extent in former ages as to induce the almost universal belief that sickness and health, prosperity and adversity, are dispensed out to each individual by Providence as the butcher dispenses meats to his customers, giving this one a good piece and that one a bad one, according to his own whim or caprice, and without any law or rule within the province of human intelligence to comprehend except the one by which the butcher is governed—that is, the amount of reward, the butcher taking his in dollars and cents while Providence should be paid in a different way. I say this has been the philosophy of the past. But, by the diffusion of intelligence and the advancement of science, much of this antiquated superstition has been dispelled, and more rational views of the nature of God's government of the world are gradually taking possession of the human mind in the more civilized and enlightened parts of the earth. As the light of science shines in upon the human mind it opens up grander views of the Divine government—views which inspire us with delight, wonder, awe and reverence. The great truth becomes conspicuously recognized, that all nature is governed by fixed and unchangeable laws, and that health and disease arise from obedience and disobedience to these natural institutions. Hence the great truth should ever be present to the mind of all, that when a child gets sick (except in those cases arising from natural causes, already spoken of), its keepers are responsible for having violated the conditions of its nature, upon which alone the Creator insures its health and safety. In other words, that a child is sick (except in the cases as above) is of itself *prima facie* evidence that it has been mismanaged, either by its parents, the public sanitary authorities, or whoever may be its keepers.

The natural laws governing the functions of the animal body are called the *organic* laws, or the *physiological* laws, or are more familiarly known as the *laws of health*. To descend into particulars and show how parents violate these laws in the management of their children would be a hopeless task in an article like this. The utmost that I can do is to point out a few of the laws and give some striking examples of their infringement.

First, may be mentioned the *dietetic* laws—those which relate to food and diet, under which head may be discussed everything that enters into the stomach. How many children are made sick unto death by violation of the dietetic laws? Mothers often begin these violations with the natal hour of their children. Ignorantly bidding defiance to the suggestions of nature, that the bland, watery fluid secreted in the mother's breast is the only substance that should enter the infant stomach, they cram it with paps, panadas, sugar, teats, candies, fruits, tea and coffee, sweetmeats, and a thousand other things, some of them very disgusting. Besides, a good assortment of wines, brandies, whiskys, soothing syrups, cordials-drops, potions, &c., is kept on hand for administration in case the little stomach should rebel against the heterogeneous incompatibles so wickedly thrust into it. I do not say that any mother gives all the articles I have indicated; but different mothers use different ones, as guests do different dishes at a hotel. I was recently called to see a woman in confinement. When I arrived the child was born and all were doing well. When I went to the bedside to salute the little stranger what was my amazement on finding it with a huge slug of old, fat, raw bacon in its mouth! I at once relieved it of the disgusting morsel, and reprimanded the mother for her ignorant conspiracy against the life of her innocent babe. She answered by saying "she thought it was right, because her mother had always served her children so." I told her that if it had been right, nature would have so arranged it that all children would have been born with a piece of bacon in their mouths. I gave orders that nothing but the mother's milk should be given it, but what was my surprise on returning two hours after to find the nurse feeding it on brandy toddy! She said " it cried as though it had the colic, and she had given it a dose of *soothing syrup*, but that did not relieve it, so she thought she must keep doing something." Think of it! an infant not six hours old had its maiden appetite spoiled by a slug of raw, fat bacon, and having failed to digest it, a dose of soothing syrup and a horn of brandy toddy were administered, and all with the very best intentions on the part of parent and nurse! The child barely escaped with its life; had it died it would have been regarded as a "mysterious dispensation of Providence." This is not an uncommon case; thousands such occur. Every experienced physician has doubtless seen numbers of similar ones. Nor are they confined to what is called the ignorant class of society.

A large majority of all the bowel complaints that afflict and destroy infancy result from violations of the dietetic laws. A child, feeble from inheritance and from the action of bad influences with which it may be surrounded, may be destroyed by a single impropriety. A bit of bread crust, a sweet cake, a nut, anything, improperly given may turn its feebleness into an actual disease which will defy all the skill of physicians and send its victim to the grave. It is from these apparently trivial things that many of the most fatal diseases arise. Like little trickling streams that issue from a thousand mountain sides in the wilderness, perhaps so small where they issue forth that a finger's end will stop the flowing veins, but as they murmur along their meandering way they unite, they gather strength from every hand, until they become a roaring torrent that defies every obstacle and sweeps everything before it in its onward course to the sea! So it is with little improprieties, little violations arising in the wilderness of ignorance, apparently trivial at first, they accumulate and gather strength as they pass on until they finally break out into a vast river of disease which bids defiance to all human power to control, as it rolls its onward way and sweeps every victim before it into the great ocean of death!

Secondly.—Violations of the *respiratory* laws are, perhaps, almost as frequent and fatal amongst children as those just considered. The respiratory laws are those governing the function of respiration or breathing. Under this head may be considered all the diseases arising from breathing an impure air. Most persons have a vague idea that it is unhealthy, especially for children, to breathe an impure air; but very few know anything more about the matter than this. What constitutes pure air and the purposes it subserves in sustaining life, and also what constitutes foul or impure air and the sources which give rise to this and the evils it engenders to health, the vast majority of the people are about as ignorant of as the man in the moon. I was recently called in haste to see an infant that had been born but an hour. The messenger informed me that it appeared very ill. On my arrival, although I had but a short distance to go, it was dead. It was full grown and to all appearance had been healthy. But the probable cause of its sudden decease was very evident on entering the room, which was small, all the doors and windows tightly closed, except a door leading into an adjoining room where the cook was preparing breakfast. The steam of hot meats and other victuals pervaded the rooms. In the lying-in room were three persons, and in the kitchen three children and the cook. The air, besides being stagnant from confinement, was so vitiated by the breath of the inmates and the exhalations from their bodies and the cook stove, that the infant perished in an hour! The parents wept over their child and marveled at the "mysterious dispensations of Providence."

I can not enumerate the sources of impure air; it would require a long chapter. It must suffice to say that tenfold more sources of vitiation arise inside of people's houses and rooms than ever enter through the doors and windows from abroad. The opprobrium of city life is "foul air;" but in my opinion it arises more from foul habits and foul rooms than any external source of vitiation. Many people are afraid to expose their children to the external air, as though it were their worst enemy. The fact is, they keep them in the house so much that when they are taken out, without great care, they take cold; just like one who has been starved for a great length of time will make himself sick by eating when furnished a sumptuous meal. Children thus kept in doors, in rooms frequently over-heated and filled with a stagnant, impure air, grow up, if they live at all, pale and sickly, like weeds grown in a dark cellar. They resemble the seeds, in the famous parable, that fell on stony ground; they soon sprang up, but, not having much root, as soon as the hot sun came they withered and died.

Too many mothers trust their children to nurses much more ignorant and careless than themselves. Many fashionable ladies are too delicate, too *nice*, now-a-days to have anything further to do with a child after it gets old enough to cry except to send for it occasionally to make an exhibition in the parlor to some *dear* friend. I observed a few days ago two little babies—ruby-lipped little jewels they were, too—sitting on the cold, stone steps of a church near my office, where they had been placed by their nurses—two "colored ladies"— who sat near by, quietly discussing their own affairs. They sat, I should think, near half an hour, for they were too young to crawl about. The day was fine over head, but all know how cold stone steps are in early spring, and how soon the cold will penetrate through several garments by sitting on them. No doubt the mothers had sent out the children into the air for their health, a thing entirely right in itself, but wrong only because they trusted them in such unsafe hands. If the children did not catch their death from the exposure it was not for want of a good opportunity. But this is only a single example; multitudes of similar ones occur continually, and keep the doctors, druggists and undertakers employed.

Woman ought to be educated in physiology and hygiene, as a part of her scholastic course.

3

Upon the mother especially devolve the duties of rearing the young. While the father is away, toiling for the means of support for the little ones, the mother is at home ministering to their wants. The mother leans upon the father for support, and the children upon the mother. To bear and rear a child aright is the most glorious work that ever fell to the lot of mortal in this mundane sphere ; and what should woman be educated for if not for this great duty ? How can she rear a child aright without any just conceptions of the nature and requirements of the infant constitution, as taught by physiology and confirmed by enlightened experience ? It is true that mothers may learn much by experience; but all will concede that experience is a dear teacher.

Home is the sphere of woman. Every element of her being proclaims her sacred to the precincts of home. Her modesty, her domestic disposition, her more delicate and polished physical structure, her insatiable love of offspring, all declare with the everlasting voice of nature that she is the tutelar deity of the household. Nor are her duties less important to the interests of mankind because thus confined to the family sphere. The home fireside is the great school-house of all nations. Upon the hearth-stone rests not only the religion, the morals and social order of every people, but likewise the pillar of State. If we would have a religious people, the nurselings must imbibe the holy principles of religion from the bosoms of their mothers. If we would have a virtuous, a moral, a courageous, a temperate people, the seeds of these precious qualities must be sown in the cradle and by the hearth-stone. If we would have a just government, administered by just, wise and good men, such men must be raised by mothers who are capable and willing to give their whole mind and soul to the glorious work. A mother, by virtue of the sacred function which makes her such, is God's vicegerent on earth, charged with the nurturing, rearing and developing a human being for life, for usefulness, for happiness, for eternity. She who underestimates the sacredness of her calling, as such, or proves recreant to its holy duties, is, to say the least of it, something less than a woman should be. She who imagines that there is a higher, more noble and useful sphere for woman than the family fireside, is lost to a true conception of the good, the right, the noble and the true.

My space is full.

My next article will be on the sources of infant mortality arising from such causes as should come within the province of public sanitary authorities.

ARTICLE V.

INFANT MORTALITY ARISING FROM SUCH SOURCES
AS SHOULD COME WITHIN THE PROVINCE OF THE
PUBLIC AUTHORITIES.

This article will be devoted to the consideration
of such sources of disease among children as
should come within the province of the public
authorities.

The reader should constantly bear in mind that
all the causes of feeblness and disease which I
have hitherto enumerated may, and often do,
operate in conjunction; that is to say, a child
may be feeble by inheritance, it may have its
vitality further impaired by improprieties of the
mother during carriage, and by mismanagement
after birth. With all these causes of enervation
operating on it the slightest exciting cause may
bring on an active disease, although it may ap-
pear in good health and may never have had an
hour's sickness in its life. A little impropriety
in diet, a change in the weather, a few hot, dry
days, breathing an atmosphere made foul by
some local cause in the neighborhood, and many
other causes acting singly or in conjunction, may
serve to excite an attack of cholera infantum,
convulsions, or some other terrible and fatal disease, any and all of which causes may have no
effect on a child that has been more favored in
constitutional vigor and previous treatment. It
is my opinion that one of the chief causes of
cholera infantum in our cities during the summer

months originates in the bad management of children during the winter. This bad management consists in keeping them too much confined in ill-ventilated, often over-heated, and frequently foul rooms. By this treatment they become so enervated that the heats of summer destroy them by this terrible disease. That it is not the action of solar heat alone that produces this disease is evidenced by the fact that it is actually less prevalent in the South than in the North. There is actually less mortality from cholera infantum in the cities of Charleston and New Orleans than in New York, Philadelphia and St. Louis—the winters being so mild in the South that children are not kept confined in doors so constantly, nor for so great a length of time, as they are during the winters at the North. Therefore they are able, in the former situation, to withstand even a greater heat with less mortality from this disease. This is a theory which I have not seen before advanced, but I believe it will furnish a key to the cause of so much cholera infantum amongst us.

I believe it will be conceded by those who have carefully read my previous articles that I have already sufficiently accounted for the great prevalence of infant mortality; nevertheless there are other causes to be considered, which, although they may be insufficient of themselves, with perhaps a few exceptions, to excite fatal diseases among children, do nevertheless, by operating in conjunction with others, no doubt prove the source of much fatality.

The chief sources of infant disease which come within the province of public authorities consist in aberrations of the atmosphere from a healthy standard, arising from local causes. This question involves the consideration of malarial influences arising from marshy soils, healthy and unhealthy sites for residences, towns, cities, etc., all of which can not be considered in so short an article. I will, therefore, confine my observations to the consideration of bad air in cities; and even this can be but brief.

The chief sources of impure air in cities are animal and vegetable substances in a state of decomposition, and the unhealthy qualities which the air acquires by its stagnation in confined places. All organic substances, that is, everything of animal and vegetable origin, as soon as life is extinct, begin to decay, to decompose, to pass back again into their original simple elements from which the life power drew them by its organizing force. During this process of decay many substances become volatilized and float about in the air as gases. These gases, when breathed or absorbed by the skin, become detrimental to health in various ways, depending on their quality and quantity. They may disturb the healthy equilibrium of the air, either by diminishing its free oxygen, by forming new compounds with some of its elements, or by simply mingling with it a foreign substance. Heat and moisture hasten the processes of decomposition and volatilization; hence it is that the air in cities is much less pure in summer than in winter.

The poisonous qualities of decaying substances can not be measured by the amount of odor they emit. The vulgar notion prevails that a decaying substance can not be detrimental unless it *stinks!* It should be borne in mind that the most deadly atmospheric poisons produce no perceptible impression on the senses. Their presence is known only by their toxical action on the system. In fact, those substances that emit an odor give off their particles in so gross a form that they seem to deteriorate the air, either by simply diluting it, so that the same volume contains less life-sustaining element, or by uniting with its oxygen, and thus generating carbonic acid and other detrimental compounds; while more deadly miasmas and contagions are so attenuated as to pervade the air without disturbing its elements, as electricity pervades gross matter.

The way to keep a city healthy is to keep it *clean!* It is not enough to keep the streets clean; they are generally kept sufficiently so for all sanitary purposes in most cities. It is in the narrow alleys, the open courts in the midst of squares, the dark, damp corners and unmentionable places that pestilence is born and bred; and it comes through the back windows, like a thief at night. People will grumble if the public authorities leave a little filth in the streets, lest it breed an unwholesome air, while the rear of their own premises is so foul that no amount of charity can plead an extenuation of their culpability. The air in the street is always moving, and as fast as effluvium is given off it is driven away by the winds. The most gentle breeze travels four or five miles an hour, while a lively gale goes from twenty to fifty in that time, and storms an hundred. People even murmur at the rudeness of the winds, whereas, without their purifying influence the inhabitants of densely populated cities would soon all perish, as the result of accumulations in the air.

There are many sources of foul air in cities which it would be tedious to mention, the most of which might be removed. St. Louis has her stock-yards and slaughter-houses inside the city limits, a thing which should not be permitted for the best of reasons.

There are some kinds of manufacturing establishments especially noxious to the health of the neighborhoods in which they are situated; among such may be mentioned soap factories, tanneries, tallow chandleries, etc., and others which emit smoke, dust and various kinds of effluvia.

We often observe slop-carts and garbage wagons passing along the streets, emiting such a foul odor that no language can describe it. Sometimes, when the air is still, in warm weather, the pestilential stench of the cart or wagon and its contents will remain for a considerable time in the track, contaminating the air and sowing the seeds of pestilence.

Vaults and water-closets, not properly drained by sewerage, is a frequent source of pestilential air.

In cities many of the poor live in old, damp, decaying buildings, in narrow alleys and in court-yards infested with vermin, rats, etc., and sur-rounded by high walls on every side, which exclude the sunlight and free air all the year round. Such places, too, are often the most re-mote from drainage and sewers. Filth accumu-lates upon filth until the very souls and bodies of such inhabitants become literally made up of the vile stuff in which they perpetually live. Such lazy, indolent people breed like spawning fishes, until each household is alive with breathing pau-pers. In many places cats, dogs, pigs, goats, etc., in great numbers, are denizened upon terms of perfect social equality with the viler biped inhab-itants.

Is it a wonder that infants die in such places ? The greater wonder is that any of them live. It is as much the duty of the public sanitary author-ities of cities to remove and prohibit such nui-sances as it is to remove the carcasses of dead animals, dung hills, etc., from the streets.

Air soon becomes unhealthy by stagnation, even as water does. Hence the great impropriety of open courts in back yards, surrounded by high walls, with no sufficient openings for the winds to drive through. Such places, although they may be enclosed in gold and silver, and kept perfectly clean, will soon become unhealthy from stagnant air; and, of course, they become more pestilential in proportion to the amount of filth they contain.

Children kept in doors during the winter, in

dark, over-heated, ill-ventilated rooms, die by the
heats of summer.   The same thing is exemplified
in many vegetables, which, if generated in warm,
moist places, excluded from the light, will perish
as soon as they are exposed to the light and heat
of the sun, unless great care be taken to make the
transition gradual.   To obviate this effect of the
summer heats on such children it seems desirable
to modify, as much as possible, the still, dry heats
of the summer months.   There are but three ways,
that I know of, to accomplish this end on a large
scale, viz.:

First.—By making the streets wider.

Secondly.—By the judicious culture of trees
and shrubs.

Thirdly.—By providing little streams of running water in every street.

All our streets are too narrow.   They ought to
be at least double their present width; and every
one, or at least every alternate one, should have
a little inclosed park, say thirty or forty feet
wide, in its midst, adorned with trees and shrub-
bery.   Tourtelle, a celebrated French physiolo-
gist, says: "The trees attract the clouds, retain
their moisture in their leaves and branches, and
are so many ventilators, which agitate and cool
the atmosphere."   They do more, they absorb
carbonic acid and give off oxygen.   It is not probable the cities of the present will be remodeled so
as to make the streets wide enough for sanitary
requirements, but future cities, built by men more
enlightened in the philosophy of man and his re-

lations to external nature, will doubtless be constructed more in harmony with these requirements. The increased facilities of locomotion which modern times have brought into use have obviated the necessity of huddling vast hordes of people in so close proximity for the sake of mere convenience. When a man can eat his breakfast at home, five or six miles out, jump into a car and be landed at his shop door in a very few minutes at a trifling expense, it is vastly better for him and society than for him to live in a narrow street, surrounded by high walls and hordes of human beings. Nay, society ought to compel him to do so for sanitary reasons, if he is not able to occupy comfortable and healthy quarters in a more central position.

The evaporation from streams of running water in the streets would do much to mitigate the excessive heat of dry summer weather. While moisture from decaying animal and vegetable matter is detrimental to health, the vapor arising from streams of running water is very beneficial in dry, hot weather. Besides cooling the air, it does much to restore and preserve the normal amount of its oxygen, which is its life-sustaining element. The heat is generally more oppressive during the first part of the night. This is due, in a great measure, to the fact that during the day the sun's rays are absorbed by the walls and pavements, and are given off—radiated—at night, so that it is often near morning before they become cooled down sufficiently to allow the air to assume

an agreeable temperature. With our present narrow streets, the heat radiated from the walls of
one side strikes the other side and serves to keep
up the oppressive warmth. The evaporation from
streams of water would not only prevent so large
an amount of heat from being absorbed during the
day, but would greatly modify the radiation at
night. A row of trees in the street would likewise moderate the heat by preventing the radia-
tions from one side passing to the other. But
trees, like every other good thing, may be abused,
and made the source of much mischief. When-
ever they become so numerous and thickly set as
to prevent the free circulation of air and the ad-
mission of sunlight, they become detrimental to
health.

My next article will treat of the causes of infant
mortality originating in the mismanagement of
physicians, and will be the closing one of my essay.

P. S.—I still hear of mutterings in the professional elements about the propriety of my articles.
The motives which actuate medical men to make
such remarks I leave for the public to judge.
Some say that "He who can write such must have
little else to do." Who wrote your text books,
Mr. M. D., from which you have derived all your
learning? Was it he who had "little else to do,"
or was it he who had dived deep into the great
arcana of nature and gathered lore from every
land? If it was the former, your books are worthless, and all your learning is "as the baseless
fabric of a vision."

ARTICLE VI.

INFANT MORTALITY ARISING FROM THE MISMANAGEMENT OF PHYSICIANS.

The subject of this article is the causes of infant mortality originating in the mismanagement of physicians.

The physician occupies a peculiarly responsible position in society; one that is not well understood nor appreciated. The patient sins against nature by disobeying the organic laws—those laws established by the Creator for the regulation and government of his physical existence. After the sin has been committed, and the patient feels the infliction of the penalty—that is, the disease—coming on him, the physician is sent for to reprieve him from the punishment. The physician thus becomes the mediator between the sinner and the law, and offers as a propitiation for the sins of his patient the means which bounteous Heaven has so munificently bestowed for the purpose, in the shape of remedies. Hence the true physician is the servant of the Divine Master, interpreting his will in the great book of nature before us, and dispensing out his blessings to many an unworthy sinner. But, in the case of the child, the parents are morally responsible for the sins committed against nature's laws, although the penalties fall upon it alone. The physician, in thus assuming the functions of an interpreter of nature and a dispenser of her blessings, often takes in his hands the reins of life and death, and should feel the high order of the responsibilities resting upon him.

From the extreme susceptibility of the infant constitution to the action of both disease and medicine, the responsibility of the physician is much greater in treating this class of patients than it is in the treatment of adults. The life-power of the infant, compared to that of the adult, is as a candle to a blazing furnace. The one may be blown out by a blast which would but serve to increase the other's heat. Hence the necessity of treating these little beings with the greatest circumspection and the mildest and gentlest medicines. The physician may fire away at his adult patient with all the artillery of the apothecary shop, day after day, and week after week, and nature will finally come out victorious and restore the patient in spite of the disease and the treatment, and the physician will get all the glory of the victory. A little anecdote will illustrate the practice of some physicians admirably. Two men were riding together along the road, one a little in advance of the other; the foremost man rode under a swinging limb of a tree; as he did so, he caught it and pulled it after him as far as he could, then letting it go, it flew back and knocked the hindmost man from his horse. He got up, badly hurt, thanked his companion, and told him that if he had not held that limb as long as he did it would have killed him. Just so do many physicians get the credit of saving their patient's lives by almost killing them. However much undeserved reputation physicians may gain by such practice among adults, it proves a blighting ruin to little children. Even those it does not send to the grave, often have implanted in their very vitals the seeds of some lasting infirmity. I have no disposition to attack any system of medicine. There is much that is good in all systems, and many good, noble and successful practitioners of them all. But it is the evils, the abuses, that I am striking at, and the results of those evils to the infant population.

The evils which are prevalent in the predominant systems of medicine may be discussed under three heads, viz.:

First.—*Heroic medication*, that is, the use of too much powerful medicine.

Secondly.—*The use of such drugs as produce lasting infirmities of constitution.*

Thirdly.—*The use of drugs in so infinitely small doses that reason and enlightened experience condemn as entirely worthless.*

The first two are peculiar to many practitioners of what is known as the "regular practice," and the last one to some of the disciples of Hahnemann.

First. — Heroic medication. Hahnemannism (homeopathy) would have died in its chrysalis state if it had not been for the wide-spread abuses of the prevailing practice. People were sick and tired of the practice that bled, salivated and blistered its victims into the grave, and eagerly caught at anything that offered the slightest shadow of a

hope of delivering them from so murderous a
practice. An hundred years ago everybody so
unfortunate as to require the services of a physician was bled, mercurialized and blistered, from
the infant in the cradle to the gray-haired sire by
his coffin. Thus seas of blood were shed, and
legions of human beings found a resting place in
a premature grave as a consequence. Yet the
doctors believed, in all sincerity and truth, that
they were doing the very utmost that human
efforts could accomplish to save their patients.
Taking every sequence as a consequence, they believed that every patient that recovered had been
saved by the treatment; and good and pious men
prayed God to bestow his blessings on their efforts
while they were unconsciously destroying their
patients! In this dark hour homeopathy was born,
and it was hailed by the people as the weary
traveler hails the flickering of a lamp in a cottage
by the wayside when he has lost his way in the
darkness of the night. It was the greatest *negative* blessing that ever dawned upon humanity
from the empire of medicine. It was as the watch-
man's lamp, it threw light upon what men were
doing in the dark. It soon demonstrated the fact
that a greater number of the sick would recover
with no medicine at all (the homeopath's being
equal to none) than did under the prevailing practice. But the people and the doctors who embraced
the new practice attributed all the good results to
the infinitesimal medicine, whereas Dame Nature
should have had all the credit. Although the rise

and progress of the HOMEOPATHIC and ECLECTIC systems of medicine have driven the lancet and many other implements of torture out of use, yet the dregs of the old theories still blubber in the bottom of the professional pot. The sticklers of these theories treat diseases as though they were killing snakes. A celebrated wit, D'Alembert, hits them by the following apologue : "Nature is fighting with a disease; a blind man armed with a club, that is, a physician, comes to settle the difference. He first tries to make peace; when he can not accomplish this he lifts his club and strikes at random. If he strikes the disease, he kills the disease ; if he strikes nature, he kills nature." Verily, such practice is not uncommon amongst us, even in the treatment of tender infants.

Secondly.—It is a well established fact that much mischief is done to children by giving them mineral poisons. If they do not prove the source of fatality at once, they often lay the foundation of lasting and incurable ills, by impairing the vital powers. They are doubtless often resorted to when, if they do not prove positively detrimental, other more efficient remedies might be used that would often save life. Calomel, calomel, calomel, is the drug constantly resorted to by some physicians, in every ailment, from the slightest ill to the gravest malady. Such physicians may be, and doubtless are, entirely conscientious : but their judgment and education are, in my opinion, sadly at fault. No drug known to the

4

human family has ever laid the foundation for so much debility of constitution, aches and pains, rotten teeth and pestilential breath as this. It is one of those drugs that wears on the axles of life, as sand on the gudgeons of a machine. The calomel doctor is the worst enemy to the future health of your family that ever set foot upon your threshold. The celebrated late Professor Nathaniel Chapman, M. D., of Philadelphia, says: *"He who resigns the fate of his patients to calomel is a vile enemy to the sick, and if he has a tolerable practice, will, in a single season, lay the foundation for a good business for life; for he will ever afterward have enough to do to stop the mercurial breaches in the constitutions of his dilapidated patients. He has thrown himself in close contact with death, and will have to fight him at arm's length so long as one of his patients maintains a miserable existence."*

Opium is a dangerous drug for infants. A half grain of Dover's powder, containing but a twentieth of a grain of opium, has been known to throw an infant into convulsions. Prof. Christison says that less than three drops of laudanum have proved fatal to stout, healthy infants. Some physicians prescribe this drug for infants with too reckless a hand. The practice so prevalent among mothers and nurses of giving laudanum, paregoric, soothing syrup, &c., is doubtless fraught with a fearful amount of death. The cause of humanity demands that these evils shall be abated.

Thirdly.—The administration of medicines in

so infinitely small doses as to produce no effect is a negative evil rather than a positive one. It, doubtless, proves the source of much fatality by allowing diseases to go on to a fatal termination that might be cured by remedies used in a more rational way. But radical homeopathy, that is, high dilutions and attenuations, is rapidly declining before the advance of rationalism; and many practitioners of that school are now using medicines in sensible doses; and some of the most successful physicians, especially in diseases of children, are among that class. Many homeopathic physicians are blatant on every available occasion in their abuses of the physicians of every other system of practice, because, as they allege, all but homeopaths give medicines in such ruinously large doses. So thoroughly have they succeeded in inflaming the prejudices of some weak-minded people in favor of their infinitesimals that many thus led astray are ready to go into tremors at the sight of an ordinary pill, although it may be made of the crumbs of bread left at their last meal; or to fall into a fit of spleen when a labeled vial of colored liquid is prepared for them, even though it may be of the same bottle of wine from which they regaled themselves at dinner. What motives are these homeopaths actuated by in thus belaboring themselves continually to frighten the ignorant, and propagate prejudice against the sanitary and rational use of medicines? It can not be philanthropy; for this is made of a different stuff. It can not be a zeal for the advancement of science and the promotion of truth, for these ends are not accomplished by such means. No; the whole secret lies in this—*their bread depends upon it.* Without such continued agitation of the subject, without such a course of carefully nursing the public prejudice, "to keep it warm," the practice could not sustain itself, nor its practioners succeed in making a livelihood by it. Different motives must actuate men who are engaged in the cause of science and the discovery and development of truth, and their application to the relief of human suffering, the advancement of civilization and the promotion of happiness.

I have before observed that the only hope of greatly reducing infant mortality lies in preventing the diseases instead of curing them. Physicians can and do achieve much, more perhaps than is generally imagined, to save life. But so long as the causes of disease are left in active operation so long will diseases arise, and many of them will prove fatal under the most enlightened and skillful management by physicians. It is true, the science of medicine is progressive; new discoveries are continually being made of means for the relief of the sick. But I doubt whether man will ever discover in nature remedies for all his physical ills. If he would avoid the penalties, he must avoid the sins.

There has ever been too much bigotry and intolerance among the medical profession. Everything new, every innovation upon the established practice, has met with the most determined opposition from the allopathic school. The priests of that school have looked for their tenets to the dim, misty superstitions of past ages, instead of searching for them in the living present and promising future. Consequently they have become learned in the rubbish of superannuated doctrines and theories instead of in the philosophy of nature and truth. Such men are bigoted, intolerant of all opposition, superficial, impractical. They strut and swell in society like great Goliaths, imagining themselves the personification and embodiment of all wisdom. A vast number of such individuals exist amongst us to-day, but they are much less numerous now than they were in former times. The rapid rise and progress, within the last few years, of the Eclectic system of medicine has done much to infuse into the profession a more liberal and democratic spirit, and to expose many of the shallow theories and destructive practices that have prevailed for so many centuries among those who are pleased to style themselves "regular physicians." The Eclectics have modified the practice in many important features: they have almost driven blood-letting out of use. They have ever condemned the use of mercury, lead, antimony, arsenic, etc., as poisonous minerals which act only as foreign substances in the body, to impair its vitality and sow the seeds of premature decay. And they are still laboring with the industry and zeal of philanthropists to induce all physicians to discard them and use new and better remedies. They have discovered and introduced many new and valuable vegetable medicines, besides new modes of using old ones in curing the sick. They have done much toward rendering medicine a pleasant sanitary restorative, instead of a terror to the sick. Their practice shows a great reduction in the death rate. According to Prof. A. Jackson Howe, M. D., of Cincinnati, one will die out of every thirty cases of sickness when left alone to nature without the aid of medicine. "In homeopathic practice one dies in every thirty-three cases treated, showing a slight improvement of that treatment over no medication. In allopathic practice one dies out of every twenty-eight or nine cases treated—a result which shows not that allopaths never administer relief, or assist in performing cures, but, *as a whole*, the fatality of that practice is a trifle more than where no medicine is given. Fifty years ago, when every disease was attacked with a heroic system of depletion, one death occurred in every twenty-five cases treated. From reports lately obtained from several hundred physicians I am warranted in stating that in the practice of eclectics *only one patient dies out of sixty cases treated.* Besides the saving of life, it is reasonable to suppose there has been a corresponding economy of valuable time and expense to those who have been sick, rendering the benefits received of still more importance."*

*An address delivered in Indianapolis, before the State Eclectic Convention, by Prof. A. Jackson Howe, M. D., Aug., 1867.

As the physician is the interpreter of nature and the dispenser of her blessings to the afflicted, it is his imperative duty to seek out the means of relief in every recess of her vast domain. He is the true physician who can lay aside all prejudice, and acknowledge truth wherever it is to be found. The man who confines himself to a set of text books, containing the written creed of a sect, and believes in them lies all the wisdom of the world, has an understanding too little to comprehend the broad domain of science or the depth of his own ignorance, and would better serve humanity at the plow than he can in the healing art. The cry quack, quack, quack! at every physician who refuses to square his opinions and practice by the tenets of the written law, or to ease his conscience at the confessional of a professional clique, has about ceased to command any respect among intelligent people. The people will judge for themselves, and give every physician credit for what he can do to restore the sick. Utility is the demand of rationalism, and rationalism is the demand of the age. The true spirit of the profession is clothed in the majesty of truth and reason; "suffereth long and is kind; it envieth not; it is not puffed up; is not easily provoked; thinketh no evil; beareth all things; hopeth all things; endureth all things, for the sake of the sacred cause in which its generous philosophy is engaged."

This ends my essay. If I have contributed anything to the dissemination of useful knowledge, I am amply paid for the labor done. If there be any whose interests I have damaged, "him have I offended."

A TREATISE

ON THE

PHYSIOLOGICAL MANAGEMENT

OF

INFANCY AND CHILDHOOD.

EMBRACING

A complete system of Hygiene for the use of parents in rearing families, with many useful suggestions on the moral and intellectual train- of childhood. By John W. Thrailkill, M. D. *St. Louis. 12mo. pp. 300.*

This work has been in course of preparation for several years, and will now soon be ready for the press. It is written in a plain, vigorous and easy style, and is designed as a popular manual on the subject for parents, and the young of both sexes, who are to take upon themselves the holy duties of parentage. The contents of the volume will

embrace everything connected with the rearing of
offspring; commencing with the parents before
conception, it follows the young being through
all its various phases of development, from conception through gestation, infancy and childhood;
setting forth, in the plainest words, the natural
laws which govern and influence the physical and
mental development of the child, thus rendering
parents able to see for themselves, by the light of
natural science, what is right and what is wrong
in the management of their children, that they
may pursue the one and avoid the other. It is
not a medical book—it does not treat of the diseases of children—but its design is to instruct
parents how to bring forth and rear up vigorous,
healthy, intellectual and moral children, such as
may be a blessing to themselves, an ornament to
society and an honor to their country. The public will be notified in due time of the appearance
of the book.